student

WORKBO

GCSE Desig… nnology Project

Graphic Produc…

Nicola Deacon

Philip Allan Updates, part of the Hodder Education Group, an Hachette Livre UK company, Market Place, Deddington, Oxfordshire OX15 0SE

Orders
Bookpoint Ltd, 130 Milton Park, Abingdon, Oxfordshire OX14 4SB
tel: 01235 827720 fax: 01235 400454
e-mail: uk.orders@bookpoint.co.uk

Lines are open 9.00 a.m.–5.00 p.m., Monday to Saturday, with a 24-hour message answering service. You can also order through the Philip Allan Updates website: www.philipallan.co.uk

ISBN 978-0-34094-639-8

Printed in Spain

Philip Allan Updates' policy is to use papers that are natural, renewable and recyclable products and made from wood grown in sustainable forests. The logging and manufacturing processes are expected to conform to the environmental regulations of the country of origin.

Introduction

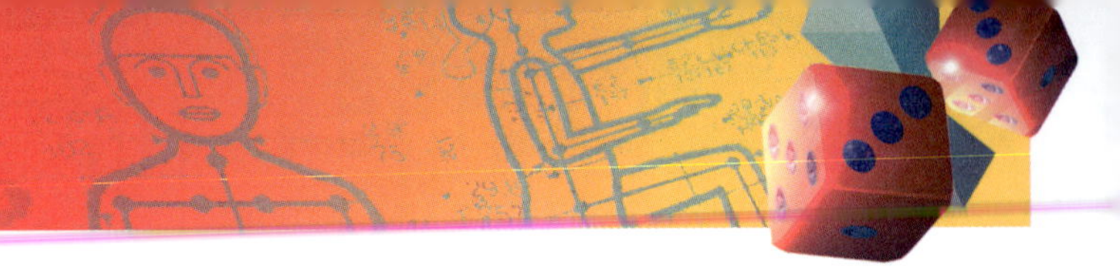

You are studying GCSE Design and Technology (Graphic Products). This means that you will specialise in paper, cardboard and perhaps some types of plastics and will learn about their properties and uses in depth. You will be assessed on one coursework project and a final examination.

The examination is worth 40% of your final GCSE grade and your coursework project is worth 60%. This workbook will help you complete your coursework project.

The project

During your course you will work on a coursework project that will probably last for up to two school terms. The project is your opportunity to show off your design and practical skills and to design and make something you are interested in. The project will be made up of a design folder and a practical product which should take you approximately 40 hours to complete.

Your project should be a product that could be made and sold commercially, so you are required to identify a design need and a target market for your proposal.

Extra marks are also awarded for the quality of your written work and your use of the technical vocabulary associated with design and technology. It pays to read your work through and check your spelling.

Assessment

The coursework project will be marked first by your teacher and then by a moderator from the awarding body. The moderator's job is to check that your teacher has marked your project to the same standard as other teachers in the country. The moderator may adjust your teacher's marks up or down.

Unlike your teacher, the moderator does not see you working or have the opportunity to speak to you about your project. All the information he or she needs to assess your performance must be present in your design folder and your practical work. The moderator should be able to see the reasons for your decisions, and how to make your product.

Exemplar student work

The illustrated design sheets included in this workbook are examples of real student work and are there to give you an idea of how the design sheets in each section might look. Although most are of a good standard, they are not meant to be right answers to coursework sheets and should not be copied. Instead, you should produce coursework sheets which are relevant to your project and show your ideas. The sheets shown cannot be given an A*–G grade as coursework projects are marked holistically or in sections for their content, not by individual sheets.

1 Presentation and communicating your ideas

Tools of the trade

Before you begin work, it is a good idea to make sure you have all the stationery and equipment you will need to complete your coursework folder. This helps you to use your time effectively both in lessons and at home. Items you will need include:

- drawing pencils — HB or softer
- colouring pencils
- pencil sharpener
- black fine liner
- plastic eraser
- 300 mm ruler
- glue stick

Methods of presentation

There are many ways to present your coursework. Most schools choose to present work in A3 flip folders, which moderators like as they are easy to look through and light to carry. Other alternatives are sketchbooks, which allow you to be creative with your presentation but which can cause problems if work is done out of order.

As technology has become an important part of school life, examination boards are recognising that many students might like to use ICT applications to present their design coursework. The examination board needs to be advised of this before you begin and will be able to give you advice and guidance about which application to use and what it expects you to produce.

Tips for presentation

A well-presented folder will make you feel proud of your work. Students who take care over their folder sheets often stay enthusiastic and keen to succeed throughout the project.

Choosing a presentation style

To help you decide on a style or 'look' for your design sheets, you could produce an image board with pictures of similar products, the environment your product would be used in and the types of people who would use it. An image board can help you come up with ideas for your presentation style. Ask questions about the pictures you have chosen. What are the most common colours? Are there any recurring shapes or motifs?

Using ICT

Consider word processing, cutting and pasting titles onto your design sheets. Find an image you can use for a motif to cut and paste onto your design sheets. If your handwriting is not very neat or if you know your spelling may need checking, word process your pieces of extended writing. You can then spell check them, print them out and stick them onto your sheets.

Borders, titles and motifs

Choose a simple style for your borders and titles, which can be reproduced easily and quickly. Marks are awarded for communicating your ideas both by hand and using ICT, so it is worth spending some time at the beginning considering your method of presentation. However, significantly more marks are given for the actual design work than for presentation, so you must make sure that you spend most of your time on design and not, for example, on colouring in a complicated border.

An alternative to a complicated border is to use a drawing or motif in the corner of each page. This should be linked to your theme: a palm tree or a pair of shoes, for example. Keep the border and titles relatively small as it is the work on the page that gains you most marks.

Reproducing your border or motif

Create a black outline of your border or a motif that can be photocopied to save time drawing it out repeatedly. You can then colour in the photocopied sheets. You could even include a background picture to give your sheets a professional feel.

Using a layout grid

Planning the layout of your sheets in advance will help you fit everything in and in an appropriate place. Create a grid on a sheet of A3 paper to put underneath your work to help you position text and pictures. You can use the grid to divide your page into columns, which will make text easier to read, especially when broken up with pictures or diagrams.

Using coloured paper

Mount designs, images and text onto coloured paper and stick them onto your design sheets. You could use a ripped or straight edge to add interest to your work.

Figure 1 Peter Rabbit has been read for generations

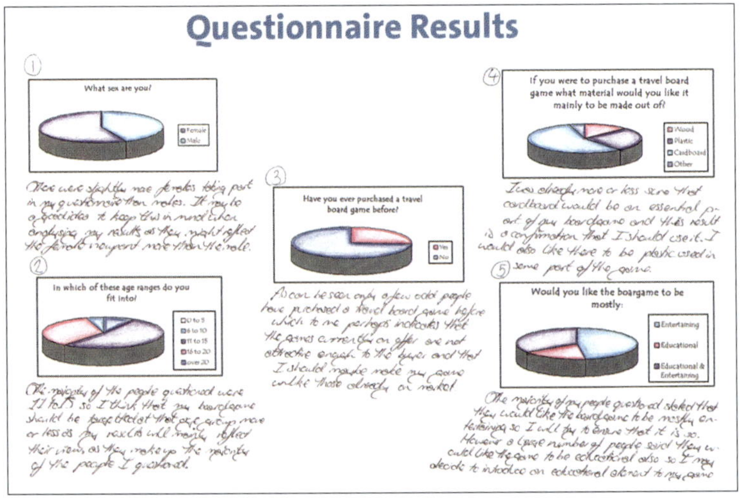

Figure 2 Example of student folder page

2 Design brief

Choosing what to design and make for your GCSE coursework project can be daunting, and many students find it difficult to get started. Your teacher may provide a brief to help you. This is a good starting point as he/she will have thought out carefully the theme and the materials and processes available at your school. The brief will help you find the best way to demonstrate your skills and abilities and ensure your project fully meets the assessment criteria.

Choosing what to design and make

When choosing what to design and make, you should:

- Make something you are interested in.
- Choose a product that satisfies a commercial need.
- Look at what is available to you in school.
- Show off your skills.
- Aim to work in a range of materials.
- Include CADCAM or write about how it could be used.
- Complete your project in 40 hours.
- Choose a realistic product that you can make in the time allowed.

Your choice of what to design and make can be the deciding factor between whether your project is a success or not. As your coursework project is worth 60% of your final GCSE grade, you should spend some time making sure you have made a good choice.

Technical demands of product manufacture

The manufacture of your product should be technically demanding. The practical making part of your project should show sufficient technical skill to have been made by a year 11 student. Your project should show your knowledge of graphical devices and layout and how to cut, fold, join and add surface decoration to materials. Some types of products do not allow you to show this level of technical skill.

Figure 1 Scrabble has proved to be a very successful graphic product

Good examples of graphic products projects include: pop-up books; activity or craft kits for children; educational, travel or board games; corporate identity and promotional items, advertisements, packaging and point of sale displays for new products.

If you choose to design the corporate identity for a new or existing company, consider carefully what products you will put the logo on (such as promotional gifts) — can you show enough technical skill in the practical making of the range of products?

Identify a buyer for your product and where it will be sold

Your project should identify a specific commercial or user need. Avoid designing for yourself (for example, for your bedroom) unless you are designing for your hobby or interest and there will be similar people interested in your product. Consider where your product might be sold and how many might need to be made if it went into commercial production. For example, if you are designing advertising or a point of sale display stand for a product, consider where it will be displayed, how much room it might take up and how much product or leaflets it will need to hold.

Incorporate 'repeatability'

To help you show your understanding of quality control and commercial production processes, you could design and manufacture a range of small modular products or one product that includes the same shape repeated several times. For example, if you choose a board game, will the playing pieces, counters or game cards be the same? Designing packaging for a product enables you to develop and use a net or template during the manufacture of your product, show your understanding of tolerances and registration when printing, and provides you with an opportunity to manufacture and test several development models.

Figure 2 Examples of graphic products

Figure 3 Example of a student product

1 Use the writing frames to explore some first ideas of what your project might be about.

1 In my spare time I like to ....

....

I am interested in designing and making a product for (activity) ....

....

....

I am interested in designing and making a product for (type of person) ....

....

....

Writing a design brief

Before you can get started, you need to identify what you are going to do and who you will be designing for. You should think about the following points:

- What type of product or product packaging could you design and make?
- Who will buy it? Who will use it? Identify the target market, their age range, sex and social background.
- Where will it be used? Identify the environment in which the product, packaging or point of sale will be used.
- Quantity manufacture. Specify that your product will be a prototype suitable to be made in commercial production. Predict how many would be made if it were to go into commercial production and why.
- Identify design constraints. What other design issues might you need to consider when designing and making your product? Will your design be sustainable? Are there any safety considerations or regulations to consider? What specific user needs might be important?

2 Complete the sentences to help you write a clear and concise design brief.

Task analysis

You may not be able to fill in all these blanks until you have given your project some more thought. Analysing the task in greater detail can help you to clarify your understanding of the design brief.

3 On a separate sheet explore your ideas for your project using mind maps (see Figure 4). Try to include everything you already know and everything you need to find out. This will help you to plan your research. Use the following topics as central themes for your mind maps:

- Target market
- Sustainable design
- Product
- Quantity manufacture
- Retail
- Safety
- Research
- Timescale
- Packaging
- Display

2 I intend to design and make a .... for .... aged between .... and .... My product will be a one-off prototype that will be suitable to be made in quantities of ....

My design will need to be ....

....

This will ensure that my product is suitable for commercial manufacture and my target market.

....

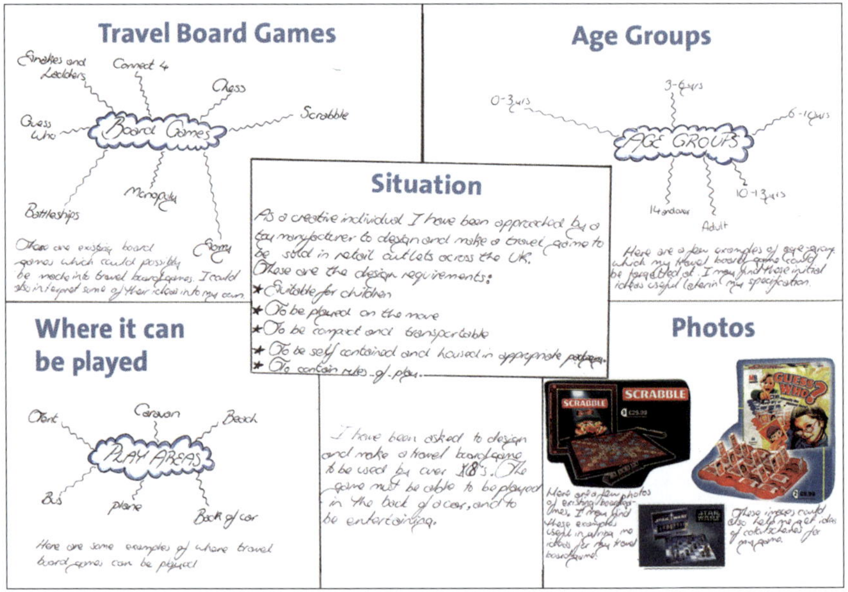

Figure 4 Example mind map

....

Planning your project

Once you have decided the type of product you are going to make, you need to plan your time carefully. Ideally you will spend approximately 40 hours on your project — this will include designing and making, so you must make sure you spend enough time on both activities. Students often spend too long designing their product and do not leave enough time to make it, losing valuable marks.

4 Ask your teacher to help you complete the project deadlines in the table.

4

Project stage	Hand-in date
1 Design brief and task analysis	
2 Research	
3 Research analysis, conclusions and specification	
4 Designing	
5 Development	
6 Planning	
7 Testing and evaluation	
8 Practical making	
9 Final hand-in deadline	

Time management

Making effective use of your time is important. At the beginning of your project the end looks far away, but before you know it, time can run out. It makes sense to plan your work to make sure you meet your coursework deadlines.

Using a Gantt Chart

A Gantt chart is a type of bar chart that shows a project schedule. The chart shows time (in weeks) across the top and lists the tasks you need to carry out down the side. You colour the squares in the chart to show how much time you intend to spend on each task and when. Planning in this way can help you to spend the right number of hours on each task and make sure everything is completed on time.

Task/week	**1**	**2**	**3**	**4**	**5**	**6**	**7**	**8**	**9**	**10**
Write design brief										
Analyse task										
Research										
Analyse research and write specification										
Designing										
Development										
Making										
Planning										
Testing										
Evaluation										

Figure 5 Example Gantt chart

5 Fill in the blanks in the sentences opposite, and then produce a Gantt chart on a separate sheet to help you plan your time. Your Gantt chart should suit your lesson and homework pattern and the deadlines set by your teacher.

5 I have .... lessons each week. Each lesson is .... minutes long. I therefore have .... minutes each week (.................... hours) to work on my project.

....

Coursework sheet

Create the design brief and task analysis design sheets for your project folder.

Present your mind maps and chosen design problem concisely, identifying what you will make, your target market, at least one design consideration and that your design will be a prototype suitable for commercial manufacture.

Present your Gantt chart and folder plan to show how you have considered what you need to do in order to design a successful solution to the design problem.

Design brief checklist

Have I...

- chosen to make something I am interested in? ☐
- chosen a project that will show off my skills and allow me to work in a range of materials? ☐
- chosen a realistic product that I can make in the time allowed? ☐
- chosen a product that will satisfy a commercial need, and identified a target market for my product? ☐
- identified how many of my product might be made if it were produced commercially? ☐
- planned how I will use my time to meet my deadlines? ☐

3 Research

What is research?

Designers and manufacturers spend a lot of time and money finding out what people need and want so they can develop products that will sell. Graphic designers work on the packaging, advertising and display to make a product more attractive to buy. Your coursework project is no different — your brief is to design and manufacture a prototype product, packaging or display solution, which would be suitable for commercial production, most likely in batch. It is important to plan what you need to find out and where you can get this information from.

Primary versus secondary research

You can find out some information 'first hand'. This is called primary research and is reliable. Other types of research (secondary research) depend on information gathered by other people and so might be less reliable. Primary research will usually gain you more credit than secondary research. However, sometimes conducting primary research is not the best use of your time — you may be able to find out the same information quicker from a secondary source.

There are many different ways to conduct your research. You need to decide which ones are most suitable for your project and are the most efficient use of your time. Students often produce pages and pages of research, which may be cut and pasted from the internet or copied from books. This type of research is not an efficient use of your time and will not gain many marks. Your research should be relevant to what you need to find out to help you design a successful product. It should show clearly and concisely why you looked for the information, what you found out and how you will use the information to help you design your product.

The research should be written in your own words, not copied. Don't forget that every student in your class has the same course book and that your teacher will have given each of your classmates the same links on the internet.

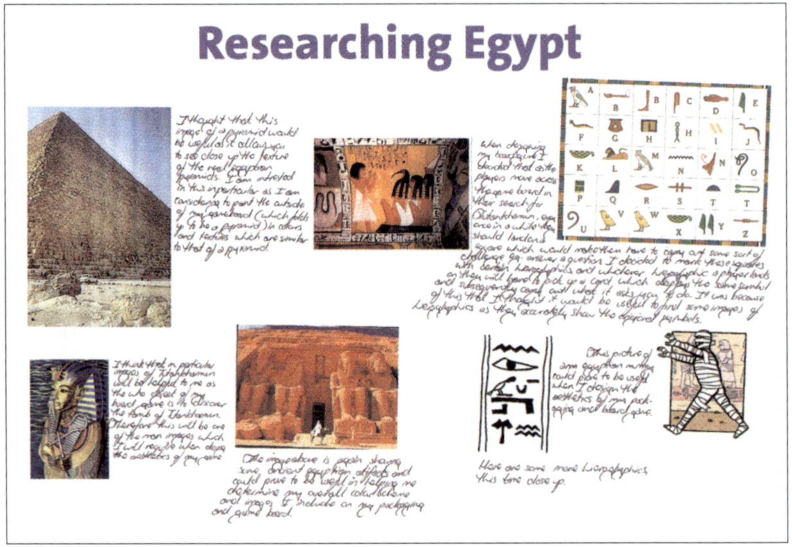

Figure 1 A student's research sheet

1 Look at the types of research listed in the table opposite. Sort them into primary and secondary types of research.

2 Look back at the mind maps you completed for the task analysis activity on page 8. Use these to help you decide what information would help you to generate ideas for a new product. On a separate sheet, create a table like the one shown in Figure 2 to plan what information you will look for, where you can look for it and what you might do with what you find.

Ideas for topics to research include:

- Target market
- Function
- Aesthetics
- Size and weight
- Human factors — anthropometrics, ergonomics, and inclusive design
- Durability and life in service
- Materials
- Manufacturing processes and quality control
- Safety regulations
- Sustainable design, product footprint and considering the environment

1

Types of research	Primary research	Secondary research
Questionnaire		
Product analysis — real example		
Product analysis — information in a catalogue		
Visit to a retail outlet		
Ergonomic and anthropometric data from a worksheet		
Photographs of products or displays taken in a local retail outlet		
Existing products, packaging and display stands		
Photographs of products found on the internet		
Printed advertisements or leaflets		

Topic	How does this relate to my project?	What do I already know?	What do I need to find out?	Where can I find the information I need?
Target market				
Size and weight				
Durability and life in service				
Aesthetics				
Human factors — ergonomics and anthropometrics, inclusive design				
Safety and regulations				
Sustainable design and considering the environment				
Materials				
Manufacturing processes and quantity				
Packaging, display and storage				
Cost				

Figure 2 Example of table used for planning research

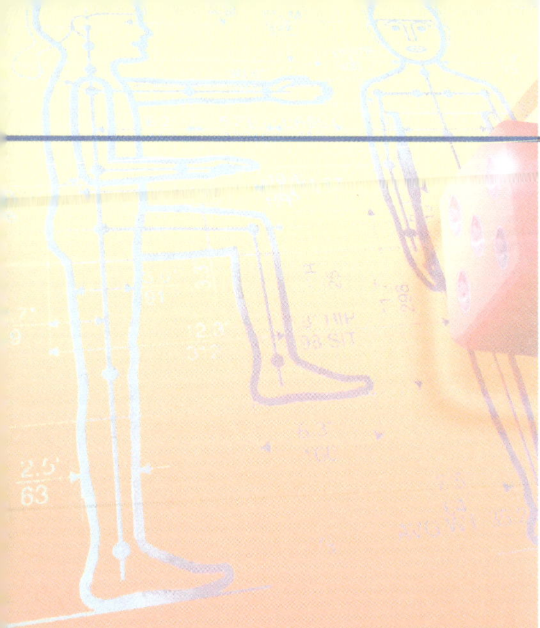

Product analysis

New graphic products are often developed as an innovative way of advertising, packaging or displaying a product to make it more attractive to buy than its competitors. Other graphic products present information in established ways such as magazines and books, but still need to attract the reader to buy them. Sometimes new technology enables products to improve, such as laser printing enabling full colour printing to be carried out cheaply.

Product designers often look at existing products to see what works well and what doesn't. This is a very effective way of developing new products, as somebody else has done most of the hard work already — the designers can build on the good parts of the design and reject the bad ones.

A good starting point for your project is therefore to look at what is already available. This product analysis is a key part of the research section of your coursework project. Product disassembly is the technique of looking at a product and deciding what processes have been used in its design and manufacture, what is good about it and what doesn't work so well.

3 Complete the sentences opposite. On a separate sheet, compare and analyse a selection of products similar to what you intend to design and manufacture. You can find information about current products by looking:

- in shops
- on the internet
- in catalogues
- in newspapers
- at product packaging
- at advertising in magazines
- at real products

3 I have decided to design and manufacture a ....

....

....

I will look for similar products to analyse by ....

....

....

....

....

....

4 a Complete the table for a product similar to what you intend to design and make.

b Choose five more examples of products similar to what you intend to design and make. On a sheet of paper create another table by writing a description of each product in the header row and filling in your ideas using the criteria below. Add two of your own criteria in the last two rows.

Criteria for product analysis:

- Ease of use
- Colour/aesthetics
- Materials
- Manufacturing processes
- Size
- Position of component parts
- Weight
- Cost
- Environmental impact
- Product maintenance

4a

Criteria	Product
Ease of use	
Colour/aesthetics	
Materials	
Manufacturing processes	
Position of component parts	
Weight	
Cost	
Environmental impact	
Product maintenance	
Other	
Other	

Creating a mood board

Designers use mood boards to help them find inspiration for new products and projects. A mood board can show shape, colour, texture, temperature, ambience and target market. 'Trendsetters' create whole books of images to show the design industry the styles and influences that will be important during future seasons. Designers use these to help them produce collections or products that people want to buy.

A mood board can help you to set out your design theme visually, and show how it might look and who might buy or use it. For example, imagine you have decided to design and manufacture an educational game for young children aged 5–10 based on the stars and planets.

Your mood board could show:

- pictures of the planets
- pictures of the night sky and star formations
- a picture of a telescope
- images of products that use the same imagery
- pictures of existing children's toys and games
- images of people typical of the target market
- similar products
- ideas for colour
- ideas for texture

Coursework sheet

Create a mood board for your product. Include images relating to colour, texture, your target market, similar products, materials and the environment where it might be used.

Look back over your research folder sheets to get some ideas.

Figure 3 A student's mood board

Figure 4 A student's mood board

Writing a questionnaire

A questionnaire is a common research method used in design coursework projects. Students often create questionnaires because they are a good source of primary research, but they might not actually help you to make any progress with your project. Before you decide to use a questionnaire, look at all the other sources of information and write a summary of what you have found out. By deciding what you already know about your potential product and your target audience, you will be able to identify what you still need to discover. Would a questionnaire help you to obtain this information? If so, you can decide who would be the best person, or people, to direct your questionnaire at, and you can write specific questions to get the answers you need (Figure 5). For example, if you are designing a new corporate identity or product packaging, you should consider who the company's customers are and ask members of that target market their opinions of your designs.

You could write a questionnaire as a group to help you interview more people. The larger the sample of opinions you collect from your target market, the more valid your results will be. If you do work as a group to collect data, you should analyse them individually. Your own interpretation of the data collected is more important than the data themselves. You should analyse the data you collect and present the results as charts or graphs (Figure 6).

1 If you were to purchase a travel game, which material would you prefer it to be made from?

Wood	☐	Plastic	☐
Card	☐	Metal	☐
Other (please specify)	☐		

____

2 Do you think a travel game should be...

Educational	☐	Entertaining	☐
Educational and entertaining	☐		

____

Figure 5 Example questionnaire

Wood	0
Plastic	3
Card	6
Metal	1
Other (please specify)	0

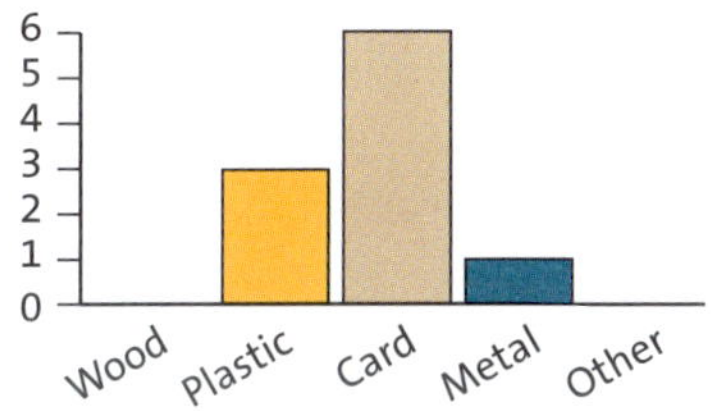

This graph shows that most people think card is the most suitable material for a travel game.

Educational	2
Entertaining	6
Educational and entertaining	2

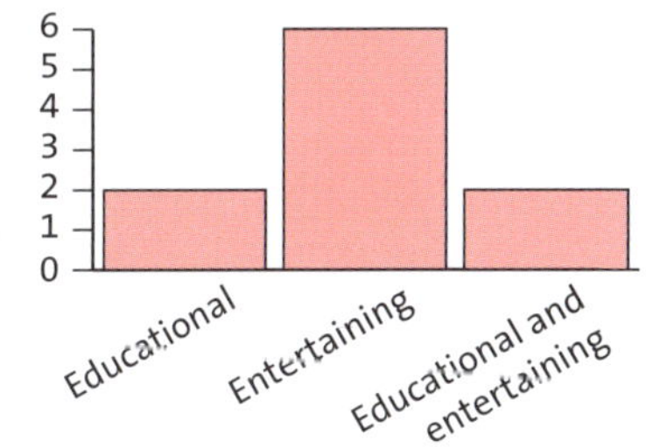

This graph shows that most of the people I asked want a travel game to be just entertaining and were not interested in any educational aspect it could have. I think that it is possible to have both, especially if the subject was fun.

Figure 6 Example questionnaire results

Presenting your research

When you have completed your research tasks, you need to decide what you should keep and what is irrelevant and can be discarded. You should also arrange your research material into a logical order and include critical analysis of the information on each page. Critical analysis is where you write what you think about the information, explain why you think it is important and how you will use it to design your product. You must talk about *your* product in detail.

Figure 7 Research can show what sort of game is likely to be popular

To gain high marks for design communication, you should use a variety of techniques to present your research. This is a good opportunity to show off your ICT skills.

Techniques you can use include:

- graphs and tables to present data collected using questionnaires and product analysis
- digital photographs of existing products
- scanned images of packaging nets or advertising leaflets
- mind maps to show how your ideas are linked
- extended writing to present information you have found in books, magazines and the internet
- mood boards to explain the aesthetic look and feel of your product
- drawings and diagrams for visual impact
- annotation: add notes, thoughts and ideas to your research to show your understanding of how it links to your product and how you might use it

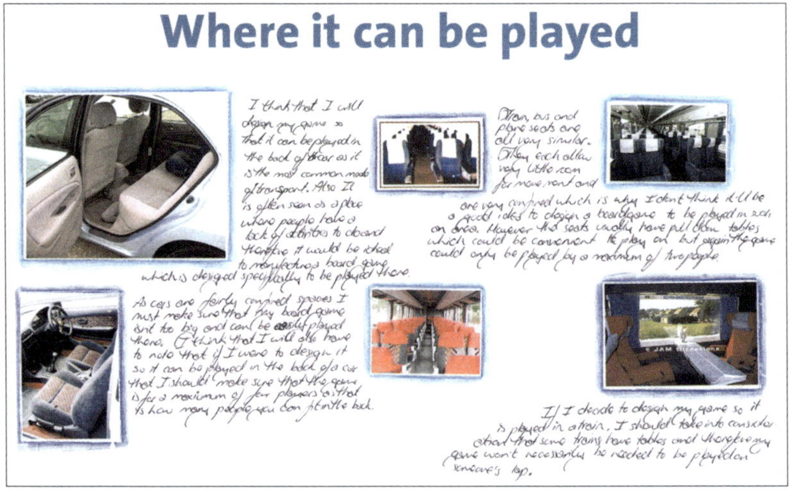

Figure 8 Example student research

5 Use the boxes to plan the content of your folder sheets in the research section of your project. Write rough notes in each box to help you decide what to put on each page.

Page 1

Page 2

Page 3

Page 4

Page 5

Coursework sheet

Create the research design sheets for your folder. Show what information you have collected and how it is relevant to your project. Remember to show the moderator that you have made choices using critical analysis.

Writing a research summary

When you have ordered and written about your research, write a summary of what you have found out and how it will help you with your designing. The summary should include a paragraph for each topic of research you have completed.

This will help the moderator to understand what you looked at and why, and how useful it was, without having to read through all your research sheets in detail. Your research summary will also help you to write a detailed specification for your design.

6 For each research topic you have completed, write a bullet point list of what you have found out. Use the lists to help you write your research conclusions. Continue on a separate sheet if necessary.

7 Use the bullet points to help you write a research conclusion on a separate sheet. This is a piece of extended writing that will give you the opportunity to show the moderator your thinking and analysis of the research you have conducted.

6 Topic: ....

I found that:

- ....
- ....
- ....
- ....
- ....

Topic: ....

I found that:

- ....
- ....
- ....
- ....
- ....

Writing a design specification

Your design specification will be a list of criteria that your product should fulfil in order to meet the needs of your target market successfully. Your research activities will help you to decide what these criteria should be.

8 Create a table like the one shown in Figure 9. This will help you write a rough draft of your design specification. You may find it helpful to divide your criteria into what your product **must** be, **should** be and **might** be, to successfully meet the consumers' needs. You should also show how each point links to your research conclusions.

Criteria	Must	Should	Might
Target market			
Function			
Size and weight			
Durability and life in service			
Aesthetics			
Human factors — ergonomics and anthropometrics, inclusive design			
Safety and regulations			
Sustainable design and considering the environment			
Materials			
Manufacturing processes and quantIty			
Packaging, display and storage			
Cost			

Figure 9 Example design specification table

Coursework sheet

Create the research conclusions and design specification sheets for your project folder. Use bullet point lists and critical analysis to summarise what you have found out and how it is relevant to your design project.

Create a detailed design specification that relates closely to your research analysis and has clear links to what you have discovered.

Research checklist

Have I...

- explained what I already know about the product I intend to design? ☐
- made sure I have used a wide variety of source types for my research? ☐
- used a mixture of primary and secondary research sources? ☐
- explained why each piece of research will help me with my project? ☐
- used critical analysis to discuss my thoughts and ideas? ☐
- found out any measurements I might need? ☐
- written a summary of what I have found out to enable me to write a thorough design specification? ☐
- linked my design specification clearly to my research and research conclusions? ☐

4 Designing

There are many techniques that can help you get started on the design of your major project. These include:

- product analysis
- improving an existing product
- comparing products
- 'in the style of...' — looking at the work of noted designers and artists
- drawing from a still life
- using nature and mathematics
- creating a mood board

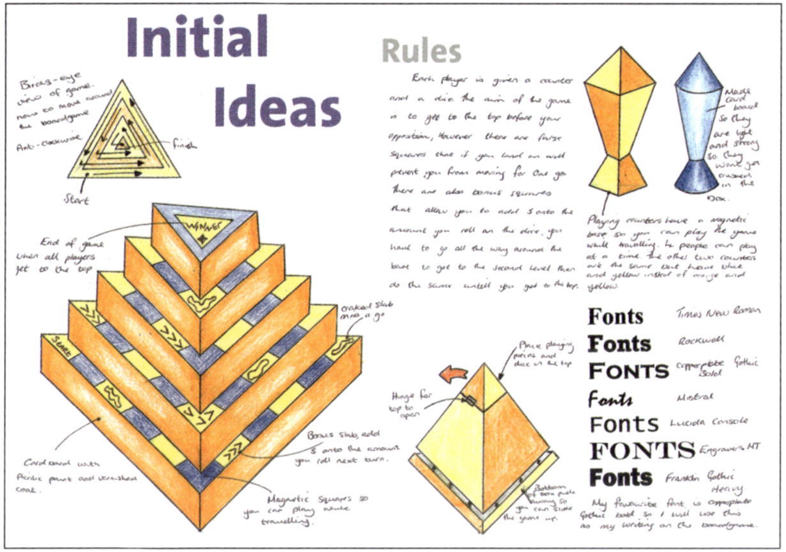

Figure 1 Example initial ideas

Design ideas

Don't think about presentation while you are creating your design ideas. Let them evolve and don't worry about how they 'sit' on the page or that they are too large or too small. Use scraps of paper or newspaper; draw upside down and side on — get those ideas out of your head and onto the page. When the ideas stop coming, you can sit back and consider what you've got. Don't discard anything. Cut or rip up these first sheets, mount your designs on coloured paper and compile them into well-thought-out and well-presented design sheets.

Annotation

Annotate your designs with your ideas and explanations about construction techniques, folding, card mechanisms, materials, colours, finishes, textures, joints, proportions, suitability for purpose and so on.

Coursework sheet

Create at least two sheets of 'thumbnail' ideas for your product. At this stage your designs should explore the concept of your product and provide details to show why each idea might differ. Don't get too hung up on the intricate details at this stage as you will have the opportunity to explore them later.

Modelling in three dimensions

Figure 2 *Example three-dimensional modelling*

Sketch modelling is a technique that creates ideas by making. Designers often use scrap materials, paper and card to help them create three-dimensional models of products as an alternative or complementary method of designing. Cutting and sticking card can help you get a feel for the size and proportion of your product and give you the inspiration to start drawing your ideas on paper. You can also take photographs of your work and include them in your coursework folder as a part of your design ideas. Another useful technique is to trace over your photographs to modify the shapes and extend the three-dimensional model using two-dimensional sketching.

Designers often produce sketch models in clay and blue foam because these materials are easy to sculpt and relatively cheap to buy. This might be helpful if you are considering vacuum forming part of your design and want to consider alternative shapes. Ask your teacher if the school has any in stock for you to try out this technique.

Using CAD to help you draw

Some students find using a computer to draw much easier than using a pencil. Working with a desktop publishing application such as Microsoft Publisher or a three-dimensional modelling application such as ProDesktop can be an effective way to generate your initial ideas. You can then print off different views of your ideas and either cut and paste them directly onto your design sheets or use them to produce renderings.

You could trace over the print-out to produce professional quality initial designs or layout grids. Use markers to add colour, highlights and shadow and a fine liner to outline your design. Use a combination of Techsoft 2D Design and Adobe Photoshop to produce packaging nets, to show your understanding of tolerances.

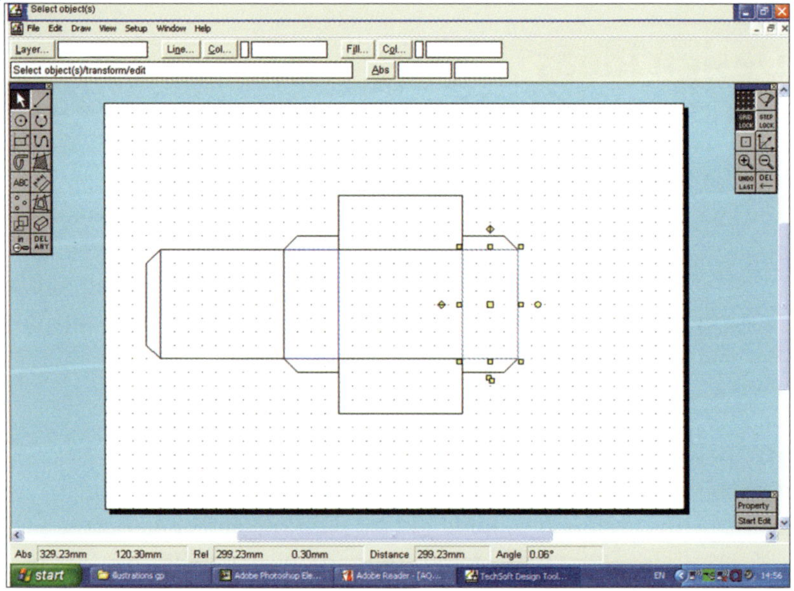

Figure 3 *Example Techsoft 2D Design net drawing*

Developing your initial ideas

It is now time to choose some of your best designs and develop them further. You may know what you want to make before you start your coursework project, but the moderator needs to see that you have considered alternative ideas and have come to a reasoned decision that your choice of product is the best solution for the design brief. It's rather like showing your working in maths — you know the answer is right, but the examiner still wants to be sure that you understand how you got there.

Choose approximately four of your most suitable design ideas and explore them further. Draw them in much greater detail, show different views, zoom in on key features and explain them with drawings and analysis. Talk about what layouts might be suitable, what colour combinations might work, alternative text and images, what pre-manufactured components could be used, surface finishes, card mechanisms and graphic materials. Consider whether you can feasibly make the product in the time available at school. Look back at your earlier annotation of your initial ideas and extend it.

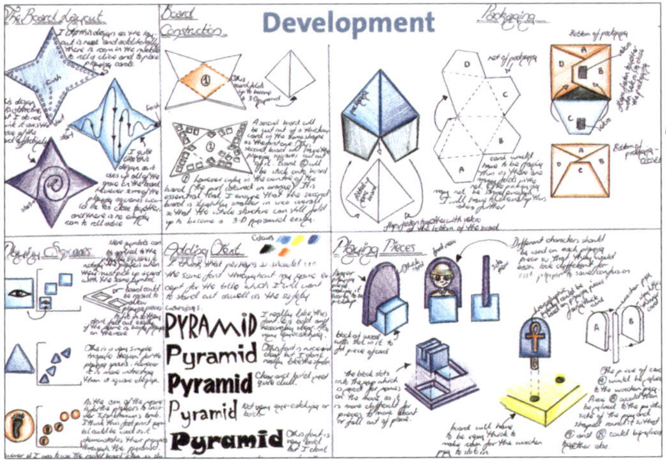

Figure 4 A student's developed design ideas

Figure 5 Trivial Pursuit is a tried and tested graphic product

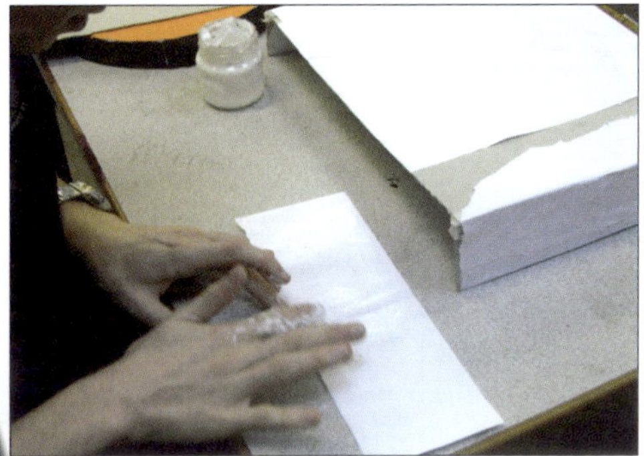
Figure 6 A student in the workshop

Coursework sheet

Create two more design sheets to develop some of your designs further. Show approximately four of your initial ideas in more detail, exploring different views, close-ups of detail, graphic layouts, alternative images, card mechanisms, graphic materials, surface finishes, colour and texture. Give alternative ideas and your opinions about which are the most appropriate.

Specification check

Include a specification check for each of your developed designs, showing how well each of your ideas meets the criteria you selected based on your research activities. Create a table like Figure 7, using the topics in your design specification to analyse your designs in a table and give them a mark out of five for each criterion.

Specification point	Design idea 1	Design idea 2
Colour/aesthetics — the travel game should be attractive to teenagers to make them want to play it.	This design uses lots of silver and gold, which reflects current teenage fashion with lots of big jewellery and chains.	This design uses very dark colours, mainly black and purple, which gives it a gothic feel.
Materials — the travel game should be made mainly from card.	This design is made entirely from card, which I think I will spray with metallic paint.	This design is made mainly from card but also uses some vacuum-formed plastic for the tray to hold the cards and counters.
Size and weight — the travel game should be small enough to be able to be taken on holiday and not take up too much space in a bag. It should also be small enough to be played on an aeroplane or in the back of a car.	This design fits into a fairly flat square box 25 cm x 25 cm. I think that it would fit into most people's travel bags although it is quite big.	This design is smaller and more likely to fit into a small bag. It would also fit onto the seat tray on an aeroplane.
Impact on the environment — the travel game should be made from sustainable materials which do not use non-renewable resources.	This design is made entirely from card so is kind to the environment, although the spray paint finish might make it less biodegradable.	The card is made from wood pulp which is a renewable resource but the plastic comes from oil which is not renewable and does not rot so will stay around for many years.
Manufacturing processes — the prototype travel game should be able to be made in the school workshop but also suitable to be made commercially in a factory.	I could make this design in a classroom and wouldn't need to be in a workshop. It could be made in multiples and would be suitable for using laminated card and to be cut using a die cutting tool.	I think this design will be easy to make in a classroom, although I would need to go into the workshop to vacuum-form the tray. Similar techniques to design 1 would be used if it were made commercially.
Cost — the travel game should cost no more than £7.50 to manufacture so that it can be sold for under £10.	I think this would be quite cheap to make as it is made from card, although metallic spray paint is quite expensive and would take up most of the budget. If I could share with another student it would bring the cost down.	This is also quite cheap, although I am not sure how much HIPs would cost — I will need to ask my teacher.
Mark out of 5	5	3

Figure 7 Example specification check

1 Create a specification check for your design. Look back at your specification and write about how well you think each of your designs meet the criteria you set. It might help you to set out your ideas in a table and to give your ideas a mark out of five like the one shown in Figure 7.

1

Decision time

Now you have analysed your design work fully, you should be able to make a reasoned decision about which is the most suitable design to take into manufacture.

Designing checklist

Have I...

- created at least two folder sheets of initial design ideas? ☐
- annotated my designs with my ideas about materials, graphic layout, surface finishes, construction techniques, colour, texture, components, measurements? ☐
- chosen at least four of my initial ideas and explored them in more detail? ☐
- used critical analysis to explain my thinking? ☐
- checked my designs against my design specification to show which is the most successful? ☐
- chosen which design I intend to manufacture? ☐

5 Developing your designs

Now that you have decided which of your designs best suits the needs of your target audience, you need to develop it into a three-dimensional prototype. A major part of this process is exploring graphic layouts, printing and surface finishes, graphic materials, and testing alternatives and developing your design further until it is the best possible.

You may need to conduct more research to help you make essential decisions about your product. You should consider what alternative materials, construction methods, methods of joining materials, components, surface decoration and finishes you might use. You should produce a folder sheet for each further research topic.

Anthropometrics and ergonomics

What measurements do you need to collect? Are there data already published or will you need to conduct some primary research? Who are you designing for? If your target users are young children aged 5 to 9 years old, you will need to consider the size of their hands, grip, reach etc. to make sure they can comfortably use your product. For example, if you are designing a point of sale display stand for children's products, you should consider where it will be used, and whether children need to be able to access it at eye/reach level. Inclusive design considers the extremes of human measurements as well as the averages to make sure everybody is able to use new products, whether they are, for example, 4 or 7 feet tall.

1 What anthropometric data do you need to help you design your product? Who will use your design? If your product is designed for an adult, would a young child be able to use it easily? Complete the statements opposite.

1 The people who will use my design are ……………………………………………

……………………………………………………………………………………

…………………………………………………………………………………… .

I will need to collect anthropometric data for people aged from ………………… to ………………… .

The measurements I will need to take are ……………………………………………

……………………………………………………………………………………

……………………………………………………………………………………

…………………………………………………………………………………… .

2 Consider how you will manufacture your design — what processes will you use? Now consider the finished product — what is it used for? Where will it be used? Who will use it?

Complete the table below to identify the properties and performance characteristics that the materials you choose will need. You may use different materials for different elements of your design. Write the elements in the first column and tick each finish, manufacturing process, property or performance characteristic that applies. Use the blank columns to add some of your own properties.

Materials, finishes, construction techniques and joints

To gain high marks for making, you should consider using more than one type of material, for example white card, paper and HIPs for vacuum-formed components. Using more than one material type will show the moderator the range of your skills and demonstrate a breadth of knowledge in graphic products. You should also explore how materials are joined together in commercial production and what construction methods are available to you in your school workshop. It is important that you show a range of technical skills to gain the highest marks — for example, making development models to test alternative construction, joining and finishing techniques.

Most graphic products are made from paper, board or plastic, but these come in infinite variations, each with different properties and performance characteristics. Some are combinations of more than one material to combine the performance characteristics of both — these are called composite materials.

	Surface						**Construction**						**Performance characteristics**							
Element of design	Paint	Print	Varnish	Laminate			Scoring	Cutting	Bending	Gluing			Strong	Greaseproof	Waterproof	Scratch resistant	Knock proof			**Possible materials**

Testing materials

You have identified the properties and performance characteristics needed by the materials you will use to manufacture your product. You can now perform tests to help you select the most appropriate materials. Some of the tests you could perform include:

- **Bending**: is the material suitable for bending or scoring? What equipment might be needed? How would this be done in industry?
- **Cutting**: what tools and equipment will be needed to cut the material? How would this be done in industry?
- **Finishing**: can the material be printed on, painted or laminated to give a commercial finish? How would this be done in industry?
- **Strength**: put the material under a heavier load than it is likely to be required to sustain in use. For example, would a point of sale stand be strong enough to hold many products?
- **Durability**: use an abrasion test to check the material's resistance to scratching; conduct an indentation test to check how resilient the material is to knocks.

Choose a variety of potential materials to test in the same way. Make sure each material receives exactly the same treatment and record your results. You could use ICT to present your findings with graphs and charts.

3 You can now test the materials you have identified to check their suitability for purpose. Complete the table opposite to show which materials you intend to compare and how you will test each one.

3

Material	Required performance characteristic	Test	Method

4 What pre-manufactured components will you need to use when making your product? Are they available from your teacher or will you need to order or buy them from a shop or catalogue? Collect information about, and prices of, the items you will need, and use it to complete the sentences opposite. Show this information in the development section of your folder.

Components

Many commercial products use pre-manufactured items in their design, as it is often more cost-effective to buy in components made elsewhere than to manufacture them on site. What pre-manufactured components could you include in your design? For example, you might need to use comb binders, split pins, screws, counters, dice, or more complex items such as electronic circuits and switches.

4 The components I will need to make my product are ....

....

....

....

I can find these ....

....

....

....

....

Industrial practices

Throughout each development topic, you should write about what might happen in industry. Commercial manufacturing techniques and methods will differ from those used in school because of the scale of manufacture and the specialist tools, equipment and processes not available to you. To gain high marks, you must show your understanding of industrial practices and how the manufacture of your product would differ if it were made commercially.

Figure 1 Vacuum forming in the school workshop

Nets and templates

Quality control is an important part of commercial production methods. Manufacturers build quality control systems into their production processes to help them reduce wastage and maintain high standards. Showing your understanding of industrial practices is an important part of your coursework project. Can you include some aspect of repeatability in the design of your prototype? Developing a net or template will help you show your awareness and consideration of quality control procedures. You could make a jig to help you line up materials for scoring or assembly or make a template for applying a decorative finish. You could explain how your net would be printed in multiples on continuous card before dies are used to cut and score it prior to assembly.

Creating a sketch model

You may wish to make several sketch models as you develop your product further. Sketch models can help you make decisions about construction techniques, joints and mechanisms. You could use paper, card, blue foam or modelling clay to test how your design looks and works, or you may wish to model only small parts of your design in the materials you intend to use to test their viability and construction techniques. You could print out images and text for your packaging and stick them onto a net to test alternative layouts. You could also print out draft-quality layouts to test registration on your net. You should photograph your test models and annotate them with your thoughts and ideas to show the moderator how your designing has progressed from your initial design to your final product.

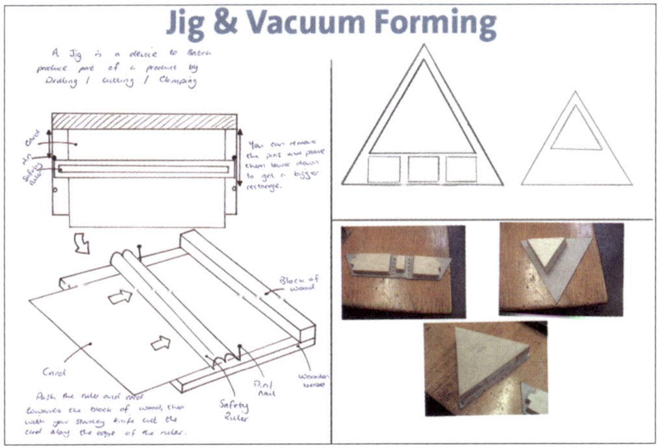

Figure 2 Example worksheet showing understanding of industrial practice

5 Which parts of your design could you test using a sketch model? Use the space on this page to plan the models you will make and the materials you will use. Use drawings and writing to help you explore your ideas.

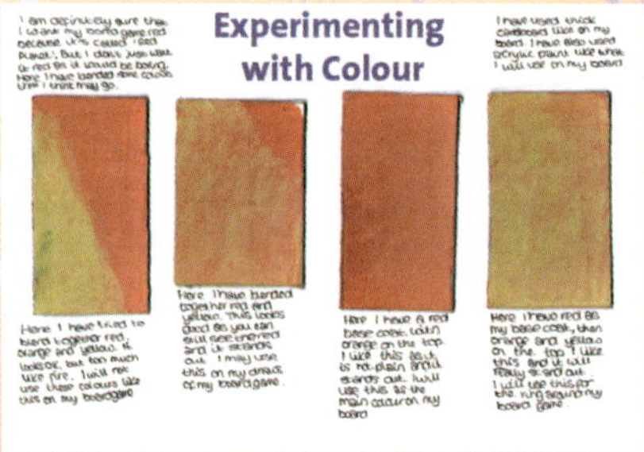

Figure 3 Student example worksheet

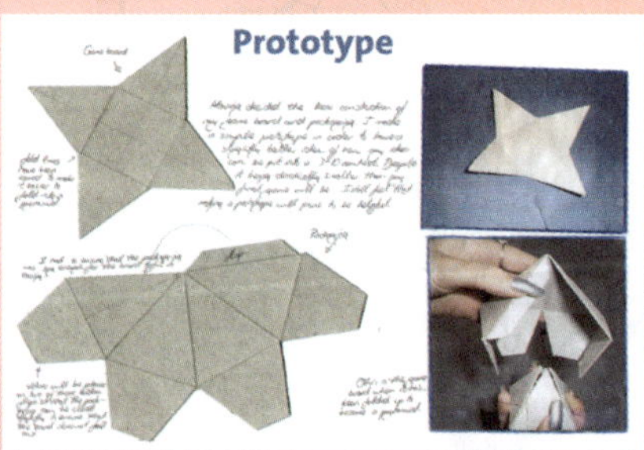

Figure 4 Student example worksheet

5

Using CADCAM

If your school has computer aided design and manufacturing facilities, you may choose to use them in the making of your product. You should test their suitability for use in the same way you would test materials and hand-finishing processes. It is not essential that you use CADCAM in your design and manufacture, as not all schools have access to the same equipment.

If you use CADCAM, you must show enough technical understanding of the process and still make sure that you spend the full time expected in the development and making of your practical product. You could compare manual techniques with CAM by testing each method and recording your experiences using a digital camera or screen dumps. You should also consider quality control and tolerances when printing, folding and joining materials. To gain high marks using CADCAM you will need to record how your designs develop over time. If you use a software application to create the surface decoration/graphic layout for your product, you should provide evidence of the time taken and your technical understanding of the process.

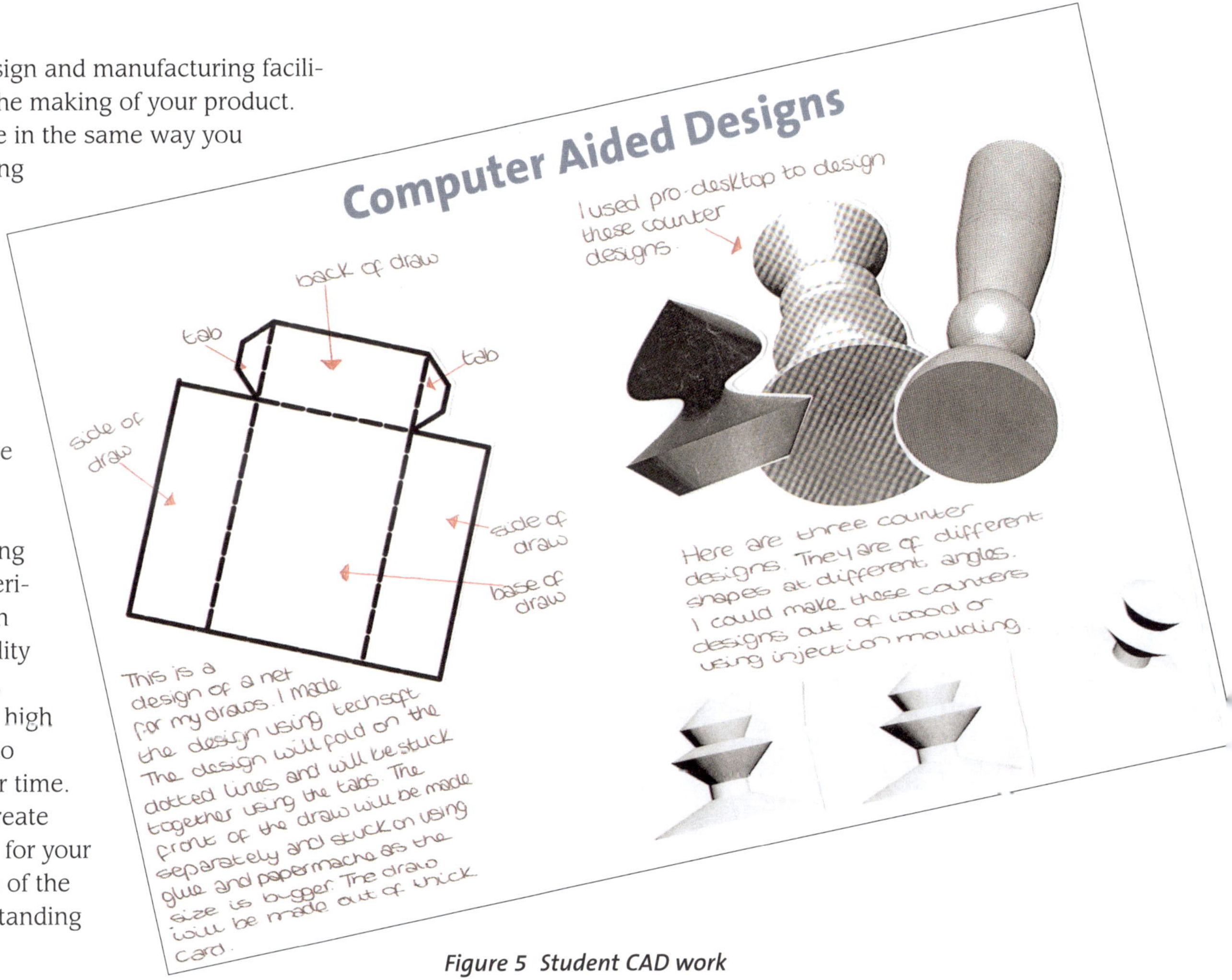

Figure 5 Student CAD work

6 Use the boxes to plan the content of your sheets in the development section of your folder. Write rough notes in each box to help you decide what to put on each page.

Page 1

Page 2

Page 3

Page 4

Coursework sheet

Create the development design sheets for your folder. Show how you have explored what materials and finishes to use and the reasons for your decisions. Photograph your sketch models and mount them on your sheets, annotating all your designs with your ideas and thoughts. Remember to use critical analysis to show the moderator why you have made particular choices.

Development checklist

Have I...

- thoroughly explored the materials and finishes I could use for my design? ☐
- identified what pre-manufactured components I will need? ☐
- designed and manufactured a net, jig or template to help with the quality control of manufacturing my design? ☐
- used critical analysis to explain my thinking? ☐
- used sketch models or draft prints to help me make decisions about how I will manufacture my design? ☐
- made sure that if I intend to use CAD and CAM to design and manufacture my design, the development and manufacture is sufficiently demanding to allow me to achieve the higher grades? ☐
- talked about how the materials, surface decoration, finishes and construction techniques I have selected for my design would be different if my design were made in industry? ☐

6 Planning

In order to meet all the assessment criteria for designing your product, you must provide detailed instructions and diagrams to explain how your product would be made. These instructions must be clear and complete so that someone else could make the product just by following them. Your instructions will be finalised after you have made your product, because you will be modifying your decisions about the best method of manufacture as you go along.

Production flowchart

A flowchart is used to divide the production into stages. This will help make your instructions easier to follow.

A production flow chart shows each stage of making. It should include quality control checks to explain how you would make sure each product was made to the same standard if more than one were made.

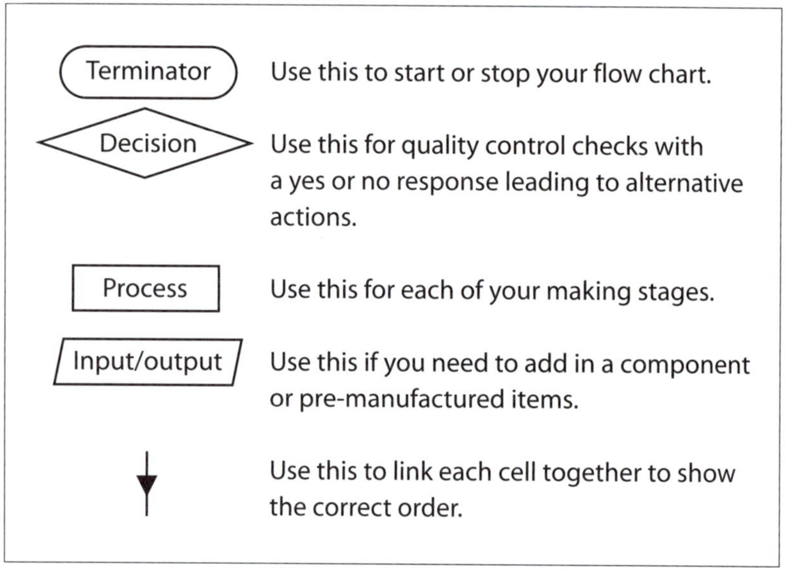

Figure 1 Symbols for production flowchart

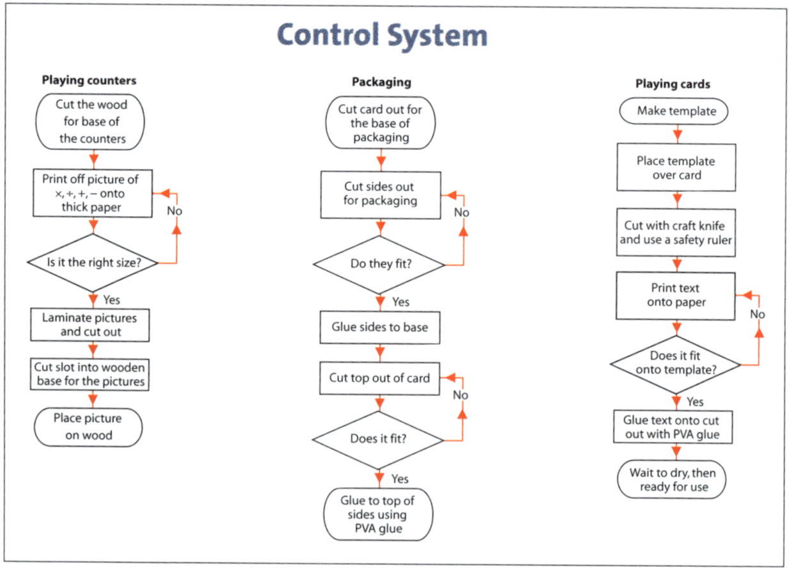

Figure 2 Example production plan flowchart

1 Plan the stages of production for your product or packaging. Write each step in the table opposite and tick which type of step it is. You may wish to complete this exercise as you go along and write up the folder sheet afterwards.

2 a On a separate sheet, create a production flowchart to show how you will make your product or packaging. Include quality control checks.

b On a separate sheet, create another production flowchart to show how you think your product or packaging would be made in a factory if it went into commercial production. Remember to think about how the order of manufacture might change, e.g. when is the card printed?

1

Stage of production	Terminator	Decision	Process	Input/ output

Coursework sheet

Create a sheet for your folder to show the two production flow charts you have made: how your product is manufactured in school and how it would be manufactured in industry.

3 On a separate sheet, create a working drawing using orthographic projection or a layout grid and a presentation drawing of your product. Remember that the information you give should enable somebody else to make your product. Your drawings should show dimensions and be to scale.

Orthographic projection

Orthographic projection shows what your product will look like from different views — front, back, left and right sides, top and bottom. You can choose to draw all the views or just the ones that are different. There is no need to draw the back if it is the same as the front, for example.

A working drawing should be to scale and show dimensions. It is a technical drawing or 'blueprint', which should be sufficiently detailed and accurate to allow a craftsman to make your product correctly.

If you are designing a leaflet or packaging, you could produce a graphic layout grid instead, showing placement of text, images and graphic devices with dimensions for margins, showing cut and fold lines and crop marks.

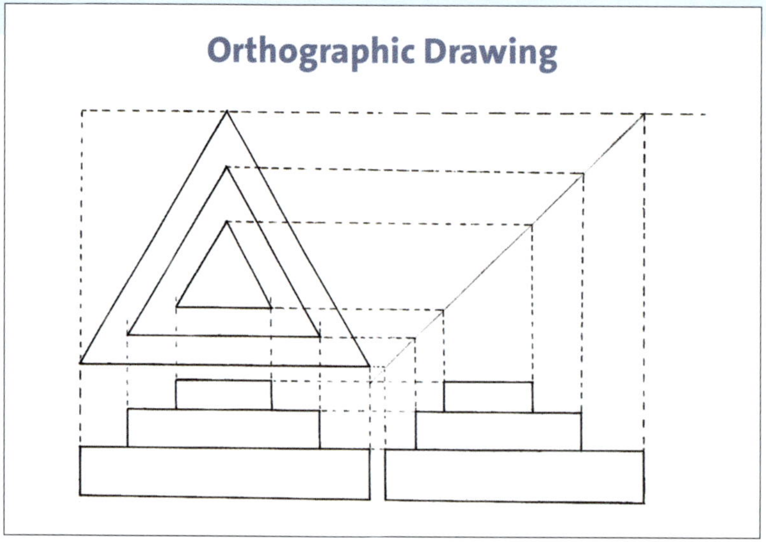

Figure 3 A student's orthographic drawing

Cutting List

Material	Component	Width	Length	Thickness	Quantity
Card	Layer of game board	390 mm	390 mm	1 mm	2
Card	Game board	390 mm	390 mm	2 mm	1
Card	Side of packaging	170 mm	170 mm	2 mm	4
Card	Closing flaps of packaging	110 mm	170 mm	2 mm	4
Card	Game board	380 mm	380 mm	1 mm	2
Card	Game board	380 mm	380 mm	2 mm	1
Wood	Playing piece	20 mm	20 mm	20 mm	3
Card	Playing piece (slotted into wooden block)	20 mm	30 mm	1 mm	3
Foam	Device to keep cards in place	10 mm	20 mm	3 mm	8

Figure 4 Example cutting list

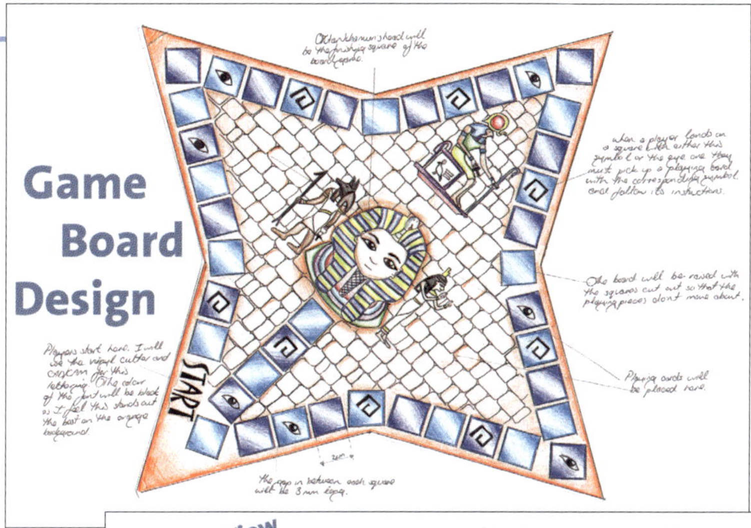

Figure 5 Example of a working drawing

Presentation drawing

Once you have completed your development activities, your final design may have changed and look different from the one in the designing section of your folder. You now need to draw your final design in detail, showing different views to enable a third party to visualise what it would look like if they followed your making instructions for themselves. If you are designing a leaflet or packaging, you should include a flat copy of your net or print of your final design before it is cut out, showing registration and crop marks.

Cutting list

As part of production planning, designers create a cutting list. This is a list of all the pieces of different materials you will need. The list includes dimensions of materials to help you cut them and the quantities and sizes of components you might need to buy.

Figure 6 A student's isometric drawing

4 Complete the table to make a cutting list of the parts and components you will need to manufacture your product.

4

Part	Size	Number required	Cost

Coursework sheet

Create a sheet for your folder to show your working and presentation drawings and your cutting list. Imagine a third party will use the information on this sheet to make your product for themselves. Have your provided enough detail?

Planning checklist

Have I...

- created a production flowchart that shows each stage needed to make my product? ☐
- considered how my production flowchart might be different if my design were made commercially? ☐
- identified the quantities and sizes of materials and pre-manufactured components I will need? ☐
- provided accurate scale drawings and visuals of my design to enable somebody else to make it? ☐

Quality of finish and attention to detail

It is now time to make your product. Your prototype should be manufactured to a commercial standard — it should be of sufficient quality to sell in a shop. It is therefore important to be accurate when measuring and joining materials together.

Being accurate and finishing your product to a high standard takes time and patience. It may mean remaking parts of your product several times or, for example, spending a long time polishing the surface. At this stage of your project, it is worth persevering with even the most frustrating practical processes. Remember that a high proportion of your coursework grade comes from your practical outcome.

Recording the accuracy checks you make and the time you take over finishing your product will show your understanding and consideration of quality control procedures.

Keeping a diary

Keeping a diary of the making part of your project can help you produce the planning section of your folder. The moderator knows that this section of your project might have been completed during or after the making of your practical product. Keeping a diary also helps the moderator understand what went into making your product and the time spent on manufacture. This is particularly useful if much of your product was manufactured using CADCAM.

You must be able to show that you spent the full time expected on the development and manufacture of your design — recording in your diary the tests you did is a good way to do this.

At the end of each lesson, write down what you have done. You should include any problems you had and how you dealt with them. If you worked with someone else, or your teacher or technician advised or helped you, note their contribution. You could draw sketches and diagrams to help you explain what you were working on, the practical work you did or the problems you encountered.

Your diary should be a visual record of the manufacturing part of your project. You could work with other people in your class to take photographs of each other working, and then paste these into your diary or folder sheets with notes to explain what you were doing. Include close-ups of specific processes and construction methods.

Figure 1 Student 'making' the design

Product Planning Action Plan

Week	Lesson	Process	Materials	Tools	Health/Safety	Quality/Control	Evaluation/ Modification
1 Board game	1 Single lesson	Cut out and measure card to length	Thick grey card	Ruler, pencil, craft knife, scissors and glue	Care needed with craft knife, use a safety ruler	Check measure-ments twice to make sure they are accurate	Could have done more that lesson
2 Board game	2 Double lesson	Masking tape board game together then glue, while drying, type up rules	Card cut from lesson 1	Glue, ruler and masking tape	Take care with craft knife. Use small amounts of glue, save time on cleaning up	Check the right amount of glue has been added	Time was planned and used well
2 Board game	3 Single lesson	Masking tape the rest of the board game together	Card cut from lesson 1	Glue, ruler and masking tape	Use small amounts of glue	Check the right amount of glue is on the board game	Went well and all glued
2 Board game	4 Double lesson	Create a papier mâché board game	Board game, paper, glue and water	Glue stick, paint brush, dish for glue and water mixture	Keep the papier mâché mixture away from mouth and eyes	Don't use too much water because it will make the board game go soggy	Was very messy with mixture all over the table
3 Packaging	5 Single lesson	Measure out packaging and cut card to length	Thick grey card	Ruler, pencil, craft knife and masking tape	Take care with the craft knife, use a safety ruler	Double check the measurements	Could have got more done, such as folder work
3 Packaging	6 Double lesson	Tape packag-ing to shape and glue together	Card cut from lesson 5	Glue and masking tape	Use small amounts of glue	Be sure to use the right amount of glue	Done all in good time, could have done more

Figure 2 Example action plan

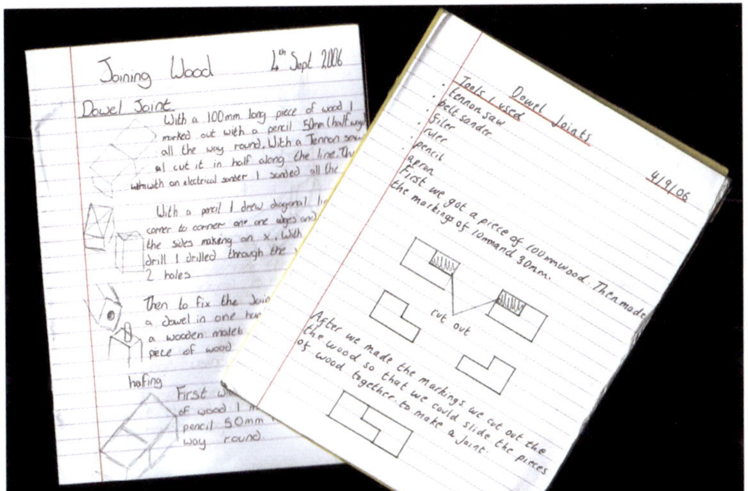

Figure 3
Example of notebook diary

Coursework sheet

Create a diary. Use either a separate notebook or draw a table on an A3 folder sheet.

Include which stage of production you are working on, the time spent and any problems you have encountered. Make a note of any help or advice you received from your teacher or technician. Include photographs of you working and close-ups of your work if you can.

Making checklist

Have I...

- spent enough time making my product? ☐
- recorded everything I did (including the problems I encountered) in my diary? ☐
- selected appropriate materials, tools and equipment and used them correctly and safely while making my product? ☐
- made sure that all aspects of my practical outcome are manufactured to a commercial standard, with attention to detail and quality of finish? ☐

8 Evaluating and testing your design

Designers in industry test and evaluate a new product at all stages of manufacture to ensure it is the best it can be and is suitable for its intended use. When you have finished making your prototype, you should evaluate how well it answers the original design brief and your design specification. You should also ask a sample of people from your target market to test your design and give their opinion.

1 Use the table to help you evaluate your prototype against your original design specification criteria. Explain why you think your design satisfies or does not satisfy the criteria and what improvements you could make.

Design specification criteria	Satisfy Y/N?	Reason	Possible modifications

2 a Create a questionnaire to help you test your prototype in the marketplace by interviewing typical customers in your identified target market group. It is helpful to prepare specific questions that would help you plan any modifications necessary for commercial production. Copy the examples opposite onto a separate sheet, and continue by creating your own.

b Interview typical people from your target market about your prototype and take photographs of them using your product. Annotate the photographs with the responses you received from your target users. If you interviewed many people, present their responses using graphs and charts. Analyse the results in a piece of extended writing.

2a What do you think about the appearance of the ....?

Do you think the colours I have used are appropriate to the person who will use the ....?

I have chosen to make the .... from .... and ....

Do you think these are appropriate materials? If not, what do you think it should be made from?

Do you like the surface finish of the ....?

Is it easy/comfortable to handle?

Is the .... too large or too small?

Do you think the .... is finished to a standard suitable to be sold in a shop?

Do you think the .... is suitable for the intended user?

Please give reasons. ....

What improvements could be made to the .... to make it more suitable?

If the .... were sold in a shop, how much would you be willing to pay for it?

Do you think the point of sale is suitable for the product/retail environment? Why do you think this?

Would the point of sale/advertisement/leaflet encourage you to buy the product? Why do you think this?

Do you think the point of sale/advertisement/leaflet/packaging would attract the target market? Why do you think this?

How does your phototype product compare with similar point of sale stands/packaging?

3 Take photographs of your finished prototype in the environment it is designed for and mount them on a folder sheet. Try to include different views so you can annotate your photographs with your ideas. Include diagrams of the modifications you might make to materials, joints or surface decoration. Note any modifications that would need to be made if your design were to be made commercially. Comment on the quality of making and explain how it could be improved. If you have manufactured packaging, you should photograph it assembled, showing different views. You should also include a flat net so that you can annotate around it with your ideas.

Improving your design and recommending modifications

After comparing your prototype with the original design brief and design specification and interviewing your target market, you may have identified some possible improvements or modifications to your design. No design is perfect, and even though you have spent much time and effort developing your prototype, another person may be able to see how you could improve some aspect of it. At this point in your project, you need to discuss what these modifications might be and show how you would incorporate them into your production plan.

Coursework sheet

Create the evaluation and testing design sheets for your folder. Show how you compared your finished prototype with the original design brief and design specification. Show any conclusions you came to and the reasons for these. Photograph different views of your prototype and mount them on your sheets, annotating them with your ideas and thoughts. Remember to show the moderator your thoughts using critical analysis.

Testing and evaluation checklist

Have I...

- compared my finished prototype with my design brief and design specification? ☐
- tested my prototype in the intended environment and with the intended user? ☐
- identified any modifications I could make to improve my product? ☐
- discussed what modifications might need to be made to my prototype if it were made commercially? ☐